I0113991

P*SSY

A Guide for Men

Lander Moore

Copyright 2016 ReagePress

ISBN Trade paperback 978-0-9963154-2-5

"I'm Pussy Galore." (Honor Blackman)
"I must be dreaming." James Bond (Sean Con-
nery) from *Goldfinger*, 1964 release

Original script dialogue
"I'm Pussy Galore."
James Bond: "I know, but what's your real name?"

Contents

~ 1 ~

Introduction

Inscrutable (def.):
impossible to understand

For men throughout the centuries and around the world, it's been inscrutable. Also unexplainable, confusing, mysterious, enigmatic, inexplicable, incomprehensible, unfathomable, unknowable …

And add baffling, bewildering ….

The woman's pussy.

The essay "10 things you didn't know about the vagina" states:

It's part of the anatomy that's shrouded in mystery, and powerful enough to cause a man to risk those things he holds dear to experience its magical wonders. Without it, life could not be given, and the sexual pleasures many now enjoy would not be known.

But for all the books, documentaries and news-paper features done on the vagina, this masterpiece—as some consider it to be—presents itself as a vexing marvel of natural engineering. Scientists are still struggling to agree on the intricacies of its design and find answers that could possibly explain the reasons behind the power it holds. With this search has come new discoveries and the debunking of myths that were once held as truth. Here are some of the facts you ought to know about the vagina, or wherever other name you choose to call it.

Key points in this essay include:

* **The vagina is part of a bigger picture.** The vagina actually refers to the canal between the vulva and the cervix, as opposed to the whole of "down there" as we are led to believe. Other parts of the female genitalia that you can expect to find "down there" also include the clitoris, the inner and outer labia, and the perineum.
* **Orgasm is a powerful painkiller.** Orgasm helps to relieve menstrual

cramps and also produces the feel good hormone that reduces feelings of depression in women. "It releases endorphins into the bloodstream, producing a sense of euphoria and leaving you with a feeling of well-being," explained Professor (Horace) Fletcher.

* **The G-spot is named after a scientist.** It took a long while for doctors to locate this elusive pleasure zone, but most if not all agree that it exists and can be found by inserting your index finger into the vagina with palm facing upwards, as you make a "come hither" motion. The G-spot is named after German gynaecologist Ernest Grafenberg, who first argued its existence in 1950.

* **The vagina is very powerful.** The vaginal walls are made of contractile muscles which makes it possible to handle the process of childbirth, for example. while many speak of the power of the vagina in making people do crazy things, the physical attributes of it cannot be overlooked. Russian born Tatyana Kozhevnikova, for example, currently holds the world record for lifting 31 pounds using only her vagina. Like any

other muscle, is physical strength can be improved with exercises such as the kegel exercise.

* **It's possible to ejaculate as a female.** We have always been told that a female cannot ejaculate, however, scientists believe it does happen, but it is often mistaken for urine during sexual intercourse. Female ejaculation was for the most part considered a myth in the past and is still considered to be so for some; but doctors have been slowly accepting the fact that it does exist. Whereas urine has a yellowish tint and comes from the bladder, this is not the case for female ejaculation.

> —All Woman section,
> www.JamaicaObserver.com,
> March 11, 2013

~ 2 ~

Names for it throughout the ages

There may be more slang names for the male and female genitalia than for anything else—ever—in the history of the world.

The website fastcodesign.com carried an article by Mark Wilson, "2,600 Slang Terms For Genitalia Throughout the Ages," with the sub-title, "You really don't want to know how your great great great great great grandmother talked."

This article summarizes the work of Jonathon Green,

> who along with his researchers, has studied more (sic) 6,000 books, dictionaries, plays, newspapers and other vintage publications dating

back hundreds of years in the interest of digging up morsels of lost speech that stemmed from the mouths of the most clever and vulgar.

Wilson also writes:

The wordplay is certainly a bit less fun when reduced to misogynistic terms, but slang for genitalia is among the most common throughout history, right alongside other topics like crime/criminals, drinking/drunks, drugs, and money. So in building out topical timelines, Green is less editorializing history … than he is mapping a subset of language evolving throughout history. In this case, history starts around 1530, when slang—words used by criminals, essentially—began getting collected. But slang exists in the written records as early as the year 1000. And in terms of these timelines, the oldest word is probably the worst … "cunt" … which made its debut around the year 1220 and has sadly stuck around.

Many of the early slang terms for the female are now virtually incomprehensible.

Following the years after 1220 are:

1367—tail.
1386—guoniam, quoint, belle-chose.
1465—socket.
1495—lap, trench.
1499—clicket-gate.
1517—token.
1520—oven.
1526—beard.
1533—mill.
1538—pudding, dock, coney, beet.
1939—leather, pin case.
1540—mawkin.
1546—jewel.
1547—plum tree.
1560—watergate.
1564—buckler.
1567—saddle.
1573—etcaetera.

And, as the years went on, more and more names—

1582—altar of Venus
1590s—bird's nest, honeypot,

treasury, fountain (of love), pitcher, Venus's cradle, Venus's glove,
chapel of ease, cuny, hell.

1600s—what-do-you-call-it, ace of
hearts, furrow.

1610s—you-know-what, engine,
phoenix nest, garden.

1630s—toy, garden of delight.

1650s—twat, lady-ware, door (labia),
cunningate, ace of clubs.

1660s—thicket, muff.

1698—first appearance of everyone's
favorite word. **Pussy**. It's now been
in use for 318 years.

~ 3 ~

The sex researchers: Alfred Kinsey Masters and Johnson

Three researchers were largely responsible for the sexual research and the sexual revolution in the twentieth century: Alfred Kinsey, William Masters and his colleague, Virginia F. Johnson.

Brief profiles in Wikipedia are worth reading:

Alfred C. Kinsey

Alfred Charles Kinsey; June 23, 1894— August 25, 1956, was an American biologist, professor of entomology and zoology, and sexologist who in 1947 founded the Institute for Sex Research at Indiana University, now known as the Kinsey Institute for Research in Sex, Gender and Reproduction. He is best known for writing *Sexual Behavior in the Human Male* (1948) and *Sexual Behavior in the Human Female* (1953), also known as the Kinsey Reports, as well as the Kinsey scale. Kinsey's research on human sexuality, foundational to the field of sexology, provoked controversy in the 1940s and 1950s. His work has influenced social and cultural values in the United States, as well as internationally.

✶ ✶ ✶

Kinsey was bisexual. He and his wife agreed that both could sleep with other people as well as each other. He himself slept with other men, including his student Clyde Martin.

✶ ✶ ✶

As a young man, Kinsey began inserting objects into his urethra—initially drinking straws before moving on to pipe cleaners, pencils and finally a toothbrush—to punish himself for having homoerotic feelings and inserting toothbrushes continued throughout his adult life. After becoming accustomed to the pain of urethral insertions, *he circumcised himself without anesthesia* (italics added).

* * *

Kinsey is widely regarded as the first public figure in American sexology; his research is cited as having paved the way for a deeper exploration into sexuality among sexologists and the general public, and as having liberated female sexuality. For example, Kinsey's work disputed the notions that women generally are not sexual and that female orgasms experienced vaginally are superior to clitoral orgasms. He initially became interested in different forms of sexual practices in 1933, after discussing the topic extensively with a colleague, Robert Kroc. Kinsey had been studying the variations in mating practices among gall wasps. During this time, he developed scale measuring sexual orientation, now known as the Kinsey scale, which ranges from 0 to 6, where 0 is exclusively heterosexual and 6 is exclusively homosexual; a rating of X for "no socio-sexual contacts or reactions: was later added.

In 1935, Kinsey delivered a lecture to a faculty discussion group at Indiana University, his first public discussion of the topic, wherein he attacked the "widespread ignorance of sexual structure and physiology" and promoted his views that "delayed marriage" (that is, delayed sexual experience) was psychologically harmful. Kinsey obtained research funding from the Rockefeller Foundation, which enabled him to further study human sexual behavior. He published *Sexual Behavior in the Human Male* in 1948, followed in 1953 by *Sexual Behavior in the Human Female*, both of which reached the top of the best-seller lists and turned Kinsey into a celebrity. These publications later became known as the Kinsey Reports. Articles about him appeared in magazines such as *Time*, *Life*, *Look* and *McCall's*. The Kinsey Reports, which led to a storm of controversy, are regarded by many as a precursor to the sexual revolution of the 1960s and 1970s.

Kinsey's research went beyond theory and interview to include observation and participation in sexual activity, sometimes involving co-workers. Some of the data published in the two *Kinsey Reports* books is controversial in the scientific and psychiatric communities, due to the low amount of research that was done and Kinsey's decision to interview and sexually experiment with volunteers who may not have been representative of the general population. Kinsey justified this sexual experimentation as being necessary to gain the confidence of his research subjects. He encouraged his staff to do likewise, and to engage in a wide range of sexual activity, to the extent they felt comfortable; he argued that this would help his interviewers understand the participants' responses. Kinsey filmed sexual acts which included co-workers in the attic of his home as part of his research; biographer Jonathan Gathorne-Hardy explains that this was done to ensure the films' secrecy, which would have caused a scandal had it become public knowledge. James H. Jones, author of *Alfred C. Kinsey: a Public/Private Life*, and British psychiatrist Theodore Dalrymple, among others, have speculated that Kinsey was driven by his own sexual needs.

Kinsey collected sexual material from around the world, which brought him to the attention of U.S. Customs when they seized pornographic films in1956; he died before this matter was resolved legally. Kinsey wrote about pre-adolescent orgasms using data in tables 30 to 34 of the male volume, which report observations of orgasms in over three-hundred children between the ages of five months and fourteen years. This information was said to have come from adults' childhood memories, or from parent or teacher observation. Kinsey said he also interviewed nine men who had sexual experiences with children or told him about the children's responses and reactions. Little attention was paid to this part of Kinsey's research at the time, but where Kinsey had gained this information began to be questioned nearly 40 years later. It was later revealed that Kinsey used data from a single pedophile and presented it as being from various sources. Kinsey had seen the need for participant confidentially and anonymity as necessary to gain "honest answers on such taboo subjects." The Kinsey Institute wrote that the data on children in tables 31-34 came from one man's journal (started in 1917) and that the events concerned pre-dated the Kinsey reports.

Jones wrote that Kinsey's sexual activity influenced his work, that he over-represented prisoners and prostitutes, classified some single people as "married," and that he included a disproportionate number of homosexual men, which may have distorted his studies. While he has been criticized for omitting African-Americans from his research, his report on the human male includes numerous references to African-American participants. Historian Vern Bullough writes that the data was later reinterpreted, excluding prisoners and data derived from an exclusively gay sample, and the results indicate that it does not appear to have skewed the data. Kinsey may have over-represented homosexuals, but Bullough considers that this may have been because homosexual behavior was stigmatized and needed to be better understood. Paul Gebhard, who was Kinsey's colleague from 1946 to 1956 and who also succeeded Kinsey as Director of the Kinsey Institute after his death, attempted to justify Kinsey's work in the 1970s by removing some of the subject data he alleged showed a bias towards homosexuality. After he recalculated the findings in Kinsey's work, he found only slight differences between the original and updated figures.

—Alfred C. Kinsey entry,
Wikipedia

Masters and Johnson

The Masters and Johnson research team composed of William H. Masters and Virginia E. Johnson pioneered research into the nature of human sexual response and the diagnosis and treatment of sexual disorders and dysfunctions from 1957 until the 1990s.

The work of Masters and Johnson began in the Department of Obstetrics and Gynecology at Washington University in St. Louis and was continued at the independent not-for-profit research institution they founded in St. Louis in 1964, originally called the Reproductive Biology Research Foundation and renamed the Masters and Johnson Institute in 1978.

In the initial phase of Masters and Johnson's studies, from 1957 until 1965 they recorded some of the first laboratory data on the anatomy and physiology of human sexual response based on direct observation of 382 woman and 312 men in what they conservatively estimated to be "10,000 complete cycles of sexual response." Their finings, particularly on the nature of female sexual arousal (for example, describing the mechanism of vaginal lubrication and debunking the earlier widely held notion that vaginal lubrication originated from the cervix) and orgasm (showing that the physiology of orgasmic response was identical whether stimulation was clitoral or vaginal, and proving that some women were capable of being multi- orgasmic), dispelled many long-standing mis-conceptions.

They jointly wrote two classic texts in the field, *Human Sexual Response* and *Human Sexual Inadequacy,* published in 1966 ands 1970, respectively. Both of these books were best-sellers and were translated into more than thirty languages. The team has been inducted into the St. Louis Walk of Fame. Additionally, they are the focus of a television project called *Masters of Sex,* for Showtime based on the 2009 biography by author Thomas Maier.

Masters and Johnson met in 1957 when William Masters hired Virginia Johnson as a research assistant to undertake a comprehensive study of human sexuality. (Masters divorced his first wife to marry Johnson in 1971. They divorced in 1992.) Previously, the study of human sexuality (sexology) had been a largely neglected area of study due to the restrictive social conventions of the time, with prostitution as a notable exception.

Alfred Kinsey and his colleagues at Indiana University had previously published two volumes on sexual behavior in the human male and female (known as the Kinsey Reports), in 948 and 1953, respectively, both of which had been revolutionary and controversial in their time. Kinsey's work however, had mainly investigated the frequency with which certain behaviors occurred in the population and was based on personal interviews not laboratory observation. In contract, Masters and Johnson set about to study the structure, psychology, and physiology of sexual behavior, though observing and measuring masturbation and sexual intercourse in the laboratory.

Initially, participants used in their experiments were prostitutes. Masters and Johnson explained that they were s socially isolated group of people, they were knowledgeable about sex and that they were willing to cooperate with the study. Of the 145 prostitutes that participated, only a select few were further evaluated for their genital anatomy and their physiological responses. In later studies, however, Masters and Johnson recruited 382 woman and 312 men from the community. The vast majority of participants were white, they had higher education levels, and most participants were married couples.

As well as recording some of the first physiological data form the human body and sex organs during sexual excitement, they also framed their finings and conclusions in language that espoused sex as a healthy and natural activity that could be enjoyed as a source of pleasure and intimacy.

The era in which their research was conducted permitted the use of methods that had not been attempted before and that have not been attempted since: "Men and women were designed as 'assigned partners' and arbitrarily paired with each other to create 'assigned couples.'"

* * *

Some sex researchers, Shere Hite in particular, have focused on understanding how individuals regard sexual experience and the meaning it holds for them. Hite had criticized Masters and Johnson's work for uncritically incorporating cultural attitudes on sexual behavior into the research; for example, her work concluded that 70 % of women who do not have orgasms though intercourse are able to achieve orgasm by masturbation. She, as well as Elisabeth Lloyd, have criticized Master and Johnsons' argument that enough clitoral stimulation to achieve orgasm should be provided by thrusting during intercourse and the inference that the failure of this is a sign of female "sexual dysfunction." While not denying that both Kinsey and Masters and Johnson have made major contributions to sexual research, she believes that people must understand the cultural and personal construction of sexual experience to make the research relevant to sexual behavior outside the laboratory. Hite's work, however, has been challenged for methodological defects.

Moreover, Masters and Johnson's research methodology has been criticized. First, Paul Robinson argues that because many of their participants were prostitutes, it is highly likely that these individuals have had more sexual experience and are also more comfortable with sex and sexuality in general. He says that one must approach these results with caution, because the participants do not represent the general population. Other researchers have argued that Masters and Johnson eliminated same-sex attracted participants when studying the human sexual response cycle, which also limits the generalizability of the results. Furthermore, Masters and Johnson have been criticized for studying sexual behaviors in the laboratory.

While they attempted to make participants as comfortable as possible in the lab by giving them a "practice session" before their behavior was recorded, critics have argued that two people engaging in sexual activity in a lab is a different experience compared to being in the privacy of one's home.

—Masters and Johnson entry,
Wikipedia

~ 4 ~

More names for it …

After the first introduction of **pussy**, in 1698, do you suppose everyone would be satisfied?

Of course not. Slang names still continued to be introduced, some newer ones also now incomprehensible. These are just a sample:

> 1700s—love's cabinet, garden of
> Venus, beauty spot.
> 1720s—lady's low toupee, (the)
> mouth that cannot bite, kitty,
> Venus's honeypot.
> 1730s—pleasure boat.
> 1740s—seat of pleasure.
> 1750s—black thing, Venus's hall,
> Venus's field, nether mouth, cock-
> pit, center of attention, central
> furrow.

1770s—Bushy Park (pubic hair), mother of all saints, mark of the beast, cradle, you-know-where, bird's nest, nest in the bush.

1775—shrine of Venus.

1780s—gash, cock alley, miraculous pitcher (that holds water with the mouth down), mantrap, snatch.

1790s—petticoat lane, Miss Brown.

1830s—Mossy cave (pubic hair), crevice, snatch-box, jewel case, fanny, blind eye (also slang for the penis), thatch (pubic hair), down below.

1840s—clam, tulip.

Green's survey shows a major jump in slang names from the 1880s to the 1910s, including: shagging machine; love crack; old lady; pleasure-pit; Lady Jane; central office; poor man's blessing; periwinkle; road to heaven and many others during this period.

1920s—beaver, fur pie and rug appear (all referring to pubic hair); bit of tail; poontang.

1950s—down there; hair pie.

Another jump in slang terms occurred in the 1960s, perhaps coinciding with the rise of

Playboy magazine and the sexual revolution. Terms appearing during this period include: man-in-the-boat (clitoris); briar patch, lawn, garden (pubic hair); passion pit; love canal; cooter; sugar walls; (or variations—velvet walls, honey walls); lulu, baby chute.

> —"2,600 Slang Terms for Genitalia
> Throughout the Ages"
> —from the website
> fastcodesign.com

Green's survey of sexual slang ends just before 2010. Presumably half the 2,600 slang terms are about the female genitalia.

One relatively modern term is *vajay-jay*—*va-jay-jay*—recently used by Oprah Winfrey—and now by many others.

See: Stephanie Rosenbloom, "What Did You Call It?" *The New York Times*, Oct. 28, 2007.

And roughly translated from the Latin, vagina means *sword holder*.

Feel free to add your own favorite names for it here:

~ 5 ~

How women think about themselves

Most women think of sexual names discretely—and far differently than men.

Mandie Williams published the article "50 Names For Your Ladyflower That Are Better Than 'Down There'" on the internet February 3, 2014. She writes:

> Some might call my hatred of "down there" irrational, but I think it is one of the single worst things to call your lady garden. Nothing takes me out of a smutty novel quicker than referring to the magical vagina as "down there." Even rap slang is better, in my humble opinion. "Down there," like the vadge in some mysterious horror show where you could

get lost if you don't have a map. Or, like you're the starry eyed virgin in a certain BDSM themed airplane novel who doesn't know the difference between a jerk with mommy issues and a dom. "Down there' reduces your business to something you can't even name, and it's tough to take pride in your bits if they don't even get their own sassy (or raunchy , or sweet, or funny) title.

Anything that inspires as much fear, lust, and admiration as the vagina deserves a more evocative name than "down there." Some are sexy, some are silly, and some might make you blush, but I am all about taking back the power in words.

—from the website thegloss.com

Her own list of names (and her comments) include:

Baby maker.
Axe wound.
Mermaid purse.
Cock holster, *star in your own XXX western*

Vadge.

C*nt, *let's take this one back, shall we?*

Twat.

Lady Flower, *like Georgia O'Keefe's.*

Poon tang, *rapper's delight.*

Love tunnel.

Devil's doorbell. *ring my*
    b-e-l-l-l-l-l-l-l

Lady bits

Privates, *for the shy bunnies.*

Muff.

Snatch.

Beaver.

Clam.

Kitty, *PG version of an old favorite.*

Muffin.

Vajayjay.

Cookie, *what's your favorite flavor?*

Box.

Cooter, *when you want to feel like a*
    hillbilly.

Toy box, *fun for hours.*

Pie, *delicious with whipped cream.*

A reader submitted the suggestion "China," she said, *"because only the best eat off my fine china."*

~ 6 ~

An anatomy lesson

Chances are, if you are not a gynecologist or a obstetrician, you know little of the female anatomy. Even women holding a hand mirror between their legs can see little of their own anatomy.

These are the major components of the female:

* The Vulva—the external opening or reproductive tract in a female.
* The Clitoris, above the vaginal opening, the "female penis" exists only for sexual pleasure, has no part in reproduction or pregnancy. (More about the clitoris, or "clit," following.)
* Labia Majora—outer (vertical) lips.
* Labia Minora—inner (vertical) lips.

* Urethra—from the bladder into the vulva, for urination.
* Vaginal opening.
* Bartholin glands—(two matching) which secrete mucus to lubricate the vagina.
* Perineum—the flesh between the vulva and the anus.

The Vulva

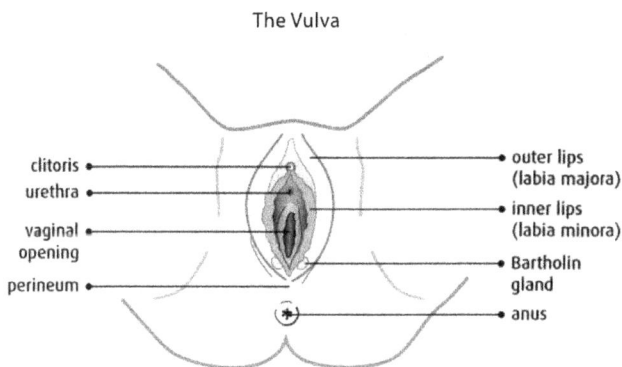

clitoris

urethra

vaginal opening

perineum

outer lips (labia majora)

inner lips (labia minora)

Bartholin gland

anus

~ 7 ~

10 Questions About Your Vagina, Answered

Women constantly, continually, have anxious questions about themselves; fortunately, there are experts ready to answer them.

In the article "10 Questions About Your Vagina, Answered," writer Jessica Bloch enlisted Lauren Streicher, associate professor of obstetrics and gynecology at Northwestern University and author of *Love Sex Again* and Hilda Hutcherson, clinical professor at the Columbia University College of Physicians and Surgeons and author of *What Your Mother Never Told You About Sex*.

Their key questions and answers included the following:

Do all vaginas look basically the same?

This question, Dr. Hutcherson says, is usually code for "Do I look normal?" Many women have at some point pulled out the ol' hand mirror to check up what's down below, but how do you know if all your parts look like everyone else's? First, let's clarify: when asking about what looks normal, most women are actually referring to their vulva, which includes the clitoris, labia and vestibule opening—all of which are outside the vagina, and guess what? There is no normal." There are tons of differences from woman to woman. Same goes for the actual internal organ—the vagina—which is different depending on a woman's age, height, sexual history and childbirth experience.

"Noses don't look the same, eye don't look the same, (and) vaginas all look very different," Dr. Streicher explains. "The differences are related to the sage of life they're in. There are differences in someone who's had children, someone who is sexually active, someone who is 6-feet tall versus 4-foot-8. There are lots of variations based on what their vagina has been through."

Can eating different foods change the way you taste?

That spicy Indian meal you at last night might set your mouth on fire. Will it eventually set your partner's mouth on fire too, after he goes down on you? Probably not, but Hutcherson notes that studies have shown that certain foods, such as melon and pineapple, can cause you to taste sweeter. On the flipside, meat-heavy diets can change the taste of vaginal secretions to be "not so great.," she adds. And here's a bummer: Alcohol, which I for many a way to lose our inhibitions in the sack, can also change the flavor of a vagina for the worse.

… and …

Can things like tampons get lost up there?

Sort of, but not really. We'll explain: Tampons, condoms, grapes (more on that in a minute) can get lost—or really, lost enough that it's out of your reach and requires a professional to pull them out.

But don't fret: "(The item) doesn't leave the vagina—there' a thought that it somehow travels up and will come out your mouth."

Streicher jokes. "Sometimes you can have something that gets lodged that isn't easily retrievable and a gynecologist can certainly do that."

* * *

Tampons and condoms are one thing. Food play during sex, however, can lead to some especially nasty vaginal issues. "One time I pulled out a grape from somebody who was playing with food during sex and forgot to remove it," Hutcherson says. "The grape caused her really bad, foul discharge. If you forget something, you've got to get it out."

—from the website
youbeauty.com

Other topics in that internet article included:

What is "normal" vaginal discharge?
Is douching once in a while really that
   bad?
Does the vagina get bigger and
   stretched out after childbirth?
What causes vagina dryness?

Is it safe to skip your period using
  birth control pills back-to-back?
Does it smell differently at different
  times of the month?
Is waxing off all my hair down there
  bad for my vagina?

~ 8 ~

10 Things You Never Knew About the Clitoris

Writing in the website Health, Kristine Thomason cites 10 key facts about the clitoris that many women—and fewer men—would know completely:

1). It's truly unique: When it comes to climaxing, "the clitoris is really, really critical," says Jim Pfaus, Ph.D. professor and sex researcher at Corcordia University in Montreal. But that's not the only thing that makes it special. The clitoris is actually the only organ in the body with the sole function of providing pleasure.

2). It's long been a mystery …

3). It's much more than meets the eye
...

4). It's got a lot of nerve. "The clitoris is the most nerve-rich part of the vulva," says Debra Herbenick, Ph.D., sexual health educator from The Kinsey Institute. The glans contains about 8,000 nerve endings making it the powerhouse of pleasure. To get some perspective, that's twice as many nerve endings as the penis. And the potential doesn't end there. This tiny erogenous zone spreads the feeling to 18,000 other nerves in the pelvis, which explains why it feels like your whole body is being taken over by your O-M-G moment.

5). Every woman's is different.

6). It's the real G-spot.

7). It's very similar to the penis. "The clitoris and the penis are somewhat mirror images of each other, just organized differently," Rebecca Chalker, Ph.D., Professor of Sexology at Pace University, and author of *The Clitoral Truth*, says.

8). ... it even gets erect. "When we talk about erection, we can't just talk about the penis," Pfaus says. "We have to talk about the clitoris. Sure, it might be less noticeable for women, but it can definitely be observed and felt. This occurs when the vestibular bulbs become engorged with blood during arousal. The blood is then trapped here until released as orgasmic spasms.

9). Size doesn't matter. Like men, women can get self-conscious about their sexy parts. Just like the penis, clits come in all shapes and sizes. And size doesn't matter for either, Chalker says.

10). It can grow with age.

—from the website health.com

~ 9 ~

12 Crazy Amazing Facts About the Clitoris

Pamela Madsen contributed this article, to *The Huffington Post*, on the internet June 17, 2015. A few of her key topics match the Kristine Thomason article, above. A summary of her topics include:

1). If you want to address the clitoris, labia and vagina together, vulva is the all-encompassing term.
2). Fifty to 75 percent of women who have climaxes (orgasms) need to have their clitoris touched (clitoral stimulation).
3). The clitoris is only partially visible to the naked eye.
4). The clitoris grows throughout a woman life.

5). The clitoris contains 8,000 deliciously sensitive nerve endings.

6). Most of us don't know that all babies have the exact same genital tissue when they are conceived. At about 12 weeks, each baby's genitalia begin to differentiate into a penis or labia.

7). Only one quarter of the clitoris is visible; the rest of it is inside the woman's body.

8). The clitoris is designed to give a woman pleasure. It has no other function.

9). There are all kinds of orgasms. Very few women are capable of achieving an orgasm without any kind of clitoral stimulation.

10). The clitoris varies in size and shape on different women.

11). People Have all kinds of nicknames for the clitoris: rosebud; joy buzzer—and others.

12). A clitoral orgasm can bring between three and 16 contractions and can last from 10 to 30 seconds.

"The word for 'Clitoris' is from the Greek 'key,'" Madsen writes, "Understanding and getting to know the clitoris may unlock your sex life forever."

~ 10 ~

The G-spot

A concise summary of the G-spot appears in the internet dictionary Wikipedia:

> The G-spot, also called the Grafenberg spot (for German gynecologist Ernest Grafenberg) is character-ized as an erogenous area of the vagina that, when simulated, may lead to strong sexual arousal, powerful orgasms and potential female ejaculation. It is typically reported to be located 5-8 cm (2-3 in) up in the front (anterior) vaginal wall, between the vaginal opening and the urethra and is a sensitive area that may be part of the female prostate.

The existence if the G-spot has not been proven; nor has the source of female ejaculation. Although the G-spot has been studied since the 1940s, disagreement persists over its existence as a distinct structure, definition and location. A 2009 British study concluded that is existence is unproven and subjective, based on questionnaires and personal experience. Other studies, using ultrasound, have found physiological evidence of the G-spot in women who report having orgasms during vaginal intercourse. It is also hypothesized that the G-spot is an extensive of the clitoris and that this is the cause of the orgasms experienced vaginally.

Sexologists and other researchers are concerned that women may consider themselves to be dysfunctional if they do not experience G-spot stimulation, and emphasize that this is not abnormal.

Theorized structure
Location

Two primary methods have been used to define and locate the G-spot as a sensitive area in the vagina: self-reported levels of arousal during stimulation, and stimulation of the G-spot leading to female ejaculation. Ultrasound technology has also been used to identify psychological differences between women and changes to the G-spot region during sexual activity.

The location of the G-spot is typically reported as being about 50 to 80mm (2 to3 in) inside the vagina, on the front wall. For some women, stimulating this area creates a more intense orgasm than clitoral stimulation. The G-spot area has been described as needing direct stimulation, such as two fingers pressed deeply into it. Attempting to stimulate the area through sexual penetration, especially in the missionary position, is difficult because of the particular angle of penetration required.

Vagina and clitoris

Women usually need direct clitoral stimulation to orgasm and G-spot stimulation may be best achieved by using both manual stimulation and vaginal penetration. Sex toys are available for G-spot stimulation. One common sex toy is the specially-designed G-spot vibrator, which is a phallus-like vibrator that has a curved tip and attempts to make G-spot stimulation easy. G-spot vibrators are made from the same materials as regular vibrators, ranging from hard plastic, rubber, silicone, jelly or any combinations of them. The level of vaginal penetration when using a G-spot vibrator depends on the woman, because women's physiology is not always the same. The effects of G-spot stimulation when using the penis or a G-spot vibrator may be enhanced by additionally stimulating other erogenous zones on a woman's body, such as the clitoris or vulva as whole. When using a G-spot vibrator this may be done by manually stimulating the clitoris, including by using the vibrator as a clitoral vibrator, or, if the vibrator is designed for it, by applying it so it stimulates the head of the clitoris, the rest of the vulva and the vagina simultaneously.

A 1981 case study reported that stimulation of the anterior vaginal wall made the area grow by fifty percent and that self-reported levels of arousal/orgasm were "deeper" when the G-sport was stimulated. Another study, in 1983, examined eleven women by palpating the entire vagina in a clockwise fashion, and reported a specific response to stimulation of the anterior vaginal wall in four of the women, concluding that the area is the G-spot. In a 1990 study, an anonymous questionnaire was distributed to 2,350 professional women in the United States and Canada with a 55 % return rate. Of these respondents, 40 % reported having fluid release (ejaculation) at the moment of orgasm, and 82 % of the women who reported the sensitive area (Grafenberg spot) also reported ejaculations with their orgasms. Several variables were associated with this perceived existence of female ejaculation.

Some research suggests that the G-spot and clitoral orgasms are of the same origin. Masters and Johnson were the first to determine that the clitoral structures surround and extend along and within the labia. Upon studying women's sexual response cycle to different stimulation, they observed that both clitoral and vaginal orgasms had the same stages of physical response, an that the majority or their subjects could only achieve clitoral orgasms, while a minority achieved vaginal orgasms. On this basis, Masters and Johnson argued that clitoral stimulation is the source of both kinds of orgasms,, reasoning that the clitoris is stimulated during penetration by friction against its hood.

Researchers at the University of L'Aquila, using ultrasonography, presented evidence that women who experience vaginal orgasms are statistically more likely to have thicker tissue in the anterior vaginal wall. The researchers believe their findings make it possible for women to have a rapid test to confirm whether or not they have a G-spot. Professor of genetic epidemiology, Tim Spector, who co-authored research questioning the existence of the G-spot and finalized it in 2009, also hypothesizes thicker tissue in the G-spot area; he states that this tissue may be part of the clitoris and is not a separate erogenous zone.

Supporting Spector's conclusion is a study published in 2005 which investigates the size of the clitoris—it suggests that clitoral tissue extends into the anterior wall of the vagina. The main researcher of the studies, Australian urologist Helen O'Connell, asserts that this interconnected relationship is the physiological explanation for the conjectured G-spot and experience of vaginal orgasms, taking into account the stimulation of the internal parts if the clitoris during vaginal penetration. While using MRI technology, O'Connell noted a direct relationship between the legs or roots of the clitoris and the erectile tissue of the "clitoral bulbs" and corpora, and the distal urethra and vagina. "The vaginal wall is, in fact, the clitoris," said O'Connell. "if you lift the skin off the vagina on the side walls, you get the bulbs of the clitoris—triangular, crescental masses of erectile tissue." O'Connell et. al., who preformed dissections on the female genitals of cadavers and used photography to map the structure of nerves in the clitoris, were already aware that the clitoris is more than just its glans and asserted in1998 that there is more erectile tissue associated with the clitoris than is generally described in anatomical textbooks. They concluded that some female have more extensive clitoral tissues and nerve than others, especially having observed this in young cadavers as compared to elderly ones. And therefore whereas the majority of females can only achieve orgasm by direct stimulation of the external parts of the clitoris, the stimulation of the more generalized tissues of the clitoris via intercourse may be sufficient for others.

French researchers Odile Bruisson and Pierre Foldes reported similar finding to those of O'Connell's. In 2008, they published the first complete 3D sonography of the stimulated clitoris, and republished it in 2009 with new research, demonstrating the ways in which erectile tissue of the clitoris engorges and surrounds the vagina. On the basis of this research, they argued that women may be able to achieve vaginal orgasm via stimulation of the G-spot because the highly enervated clitoris is pulled closely to the anterior wall of the vagina when the woman is sexually aroused and during vaginal penetration. They assert that since the front wall of the vagina is inextricably linked with the internal parts of the clitoris, stimulating the vagina without activating the clitoris may be next to impossible. In their 2009 published study, the "coronal planes during perineal contraction and finger penetration demonstrated a close relationship between the root of the clitoris and the anterior vaginal wall." Bruisson and Foldes suggested "that the special sensitivity of the lower anterior vaginal wall could be explained by pressure and movement of clitoris' root during a vaginal penetration and subsequent pineal contraction."

Female prostate

In 2001, the Federative Committee on Anatomical Terminology accepted *female prostate* as an accurate term for the Skene's gland, which is believed to be found in the G-spot area along the walls of the urethra. The male prostate is biologically homologous to the Skene's gland; it has been unofficially called the male G-spot because it can also be used as an erogenous zone.

Regnier de Graaf, in 1672, observed that the secretions (female ejaculation) by the erogenous zone in the vagina lubricate "in agreeable fashion during coitus." Modern scientific hypothesis linking G-spot sensitivity with female ejaculation led to the idea that non-urine fame ejaculation may originate from the Skene's gland, with the Skene's gland and male prostate acting similarly in terms of prostate-specific antigen and prostate-specific acid phosphatase studies, which led to a trend calling the Skene's glands the female prostate. Additionally, the enzyme PDE5 (involved with erectile dysfunction) has additionally been associated with the G-spot area. Because of these factors, it has been argued that the G-spot is a system of glands and ducts located within the anterior (front) wall of the vagina. A similar approach has linked the G-spot with the urethral sponge.

Clinical Significance

G-spot amplification (also called G-spot augmentation or the G-shot) is a procedure intended to temporarily increase pleasure in sexually active women with normal sexual function, focusing on increasing the size and sensitivity of the G-spot. G-spot amplification is performed by attempting to locate the G-spot and noting measurements for future reference. After numbing the areas with a local anesthetic, human engineered collagen is then injected directly under the mucosa in the area the G-spot is concluded to be in.

A position paper published by the American College of Obstetricians and Gynecologists in 2007 warns that thee is no valid medical reason to perform he procedure, which is not considered routine or accepted by the college; and it has not been proven to be safe or effective. The potential risks include sexual dysfunction, infection, altered sensation, dyspareunia, adhesions and scarring. The College position is that it is untenable to recommend the procedure. The procedure is also not approved by the Food and Drug Administration or the American Medical Association, and no peer-reviewed studies have been accepted to account for either safety or effectiveness of this treatment.

Society and Culture
General skepticism

In addition to general skepticism among gynecologists, sexologists and other researchers that the G-spot exists, a team at King's College London in late 2009 suggested that its existence is subjective. They acquired the largest sample size of women to date— 1,800—who are pairs of twins, and found that the twins did not report a similar-spot in their questionnaires. The research, headed by Tim Spencer, documents a 15-year study of the twins, identical and non-identical. Identical twins share genes, while non-identical pairs share 50% of theirs. According to the researchers, if one identical twin reported having a G-spot, it was more likely that the other one would too, but this pattern did not materialize. Study co-author Andrea Burri believes: "It is rather irresponsible to claim, the existence of an entity that has never been proven and pressurize women and men too."She stated that one of the reasons for the research was to remove feelings of "inadequacy or underachievement" for women who feared they lacked a G-spot. Researcher Beverly Whipple dismissed the findings, commenting that twins have different sexual partners and techniques. And that the study did not properly account for lesbian or bisexual women.

Petra Boynton, a British scientist who has written extensively on the G-spot debate, is also concerned about the promotion of the G-spot leading women to f eel "dysfunctional" if they do not experience it. "We're all different. Some women will have a certain area within the vagina which will be very sensitive, and some won't—but they won't necessarily be in the area called the G-spot," she stated. "If a woman spends all her time worrying about whether she is normal, or has a G-spot or not, she will focus on just one area, and ignore everything else. It's telling people that there is single, best way to have sex, which isn't the right thing to do."

Nerve Endings

G-spot investigators are criticized for giving too much credence to anecdotal evidence, and for questionable investigative methods,, for instance, the studies which have yielded positive evidence for a precisely located G-spot involve small participant samples. While the existence of a greater concentration of nerve endings at the lower third (near the entrance) of the vagina is commonly cited, some scientific examinations of vaginal wall innervation have shown no single area with a greater density of nerve endings.

Several researchers also consider the connection between the Skene's gland and the G spot to be weak. The urethral sponge, however, whish is also hypothesized as the G-spot, contains sensitive nerve endings and erective tissue. Sensitivity is not determined by neuron density alone: other factors include the branching patterns of neuron terminals and cross or collateral innervation of neurons. While G-spot opponents argue that because there are very few tactile nerve endings in the vagina and that therefore the G-pot cannot exist, G-spot proponents argue that vaginal orgasms rely on pressure sensitive nerves.

Clitoral and other anatomical debates

The G-spot having an anatomical relationship with the clitoris has been challenged
by Vincenzo Puppo, who, while agreeing that
the clitoris is the center of female sexual pleasure, disagrees with Helen O'Connell and
other researchers' terminological and anatomical descriptions of the clitoris. He stated,
"Clitoral bulbs is an incorrect term from an
embryological and anatomical viewpoint in
fact, the bulbs do not develop from the phal-
lus, and they do not belong to the clitoris." He
says that *clitoral bulbs* "is not the correct term
used in human anatomy" and that *vestibular
bulbs* is the correct term, adding that gynecologists and sexual experts should inform
the public with facts instead of hypotheses or
personal opinions. "{C}litorial/vaginal/uter-
ine orgasm C/A/C/U spot orgasm, and female
ejaculation, are terms that should not be used
by sexologists, women and mass media," he
said, further commenting that the "anterior
wall is separated from the posterior urethral
wall by the urethrovaginal septum (its thick-
ness is 10-12 mm)" and that the "inner
clitoris" does not exist. "The female perineal

urethra, which is located in front of the anterior vaginal wall, is about one centimeter in length and the G-spot is located in the pelvic wall of the urethra, 2-3 cm into the vagina." Puppo stated. He believes that the penis cannot come into contact with the congregation of multiple nerves/veins situated until the angle of the clitoris, detailed by Georg Ludwig Kobelt, or with the roots of the clitoris, which does not have sensory receptors or erogenous sensitivity, during vaginal intercourse. He did, however, dismiss the orgasmic definition of the G-spot that emerged after Ernest Grafenberg, stating that "there is no anatomical evidence of the vaginal orgasm which was invented by Freud in 1905, without any scientific basis."

Puppo's belief that there is no anatomical relationship between the vagina and clitoris is contracted by the general belief among researchers that vaginal orgasm are the results of clitoral stimulation; they maintain that clitoral tissue extends, or is at least likely stimulated by the clitoral bulbs, even in the area most commonly reported to be the G-spot. "My view is that the G-spot is really just an extension of the clitoris on the inside of the vagina, analogous to the base of the male penis," said researcher Amichai Kilchevsky. Because female fetal development is the "default" direction of fetal development in the absence of substantial exposure to make hormones and therefore the penis is essential a clitoris enlarged by such hormones. Kilchevsky believes that there is no evolutionary reason why females would have two separate structures capable of producing orgasms and blames the porn industry and "G-spot promoters" for "encouraging the myth" of a distinct G-spot.

The general difficulty of achieving vaginal orgasms, which is a predicament that is likely due to nature easing the process of child bearing by drastically reducing the number of vaginal nerve endings, challenge arguments that vaginal orgasms help encourage sexual intercourse in order to facilitate reproduction. O'Connell stated that focusing on the G-spot to the exclusion of the rest of a woman's body is "a bit like stimulating a guy's testicles without touching the penis and expecting an orgasm to occur just because love is present." She stated that it "is best to think of the clitoris, urethra, and vagina as one unit because they are intimately related." Ian Kerner stated that the G-spot may be "nothing more than the roots of the clitoris crisscrossing the urethral sponge."

A Rutgers University study, published in 2011, was the first to map the female genitals onto the sensory portion of the brain, and supports the possibility of a distinct G-spot. When the research team asked several women to stimulate themselves in a functional magnetic resonance (MRI) machine, brain scans showed stimulating the clitoris, vagina and cervix lit up distinct areas of the women's sensory cortex, which means the brain registered distinct feelings between stimulating the clitoris, cervix and the vaginal wall—where the G-spot is reported to be. "I think that the bulk of the evidence shows the G-spot is not a particular thing," stated Barry Komisaruk, head of the research findings. "It's not like saying 'what is the thyroid gland?' The G-spot is more of a thing like New York City is a thing. It's a region, it's a convergence of many different structures."

In 2009, *The Journal of Sexual Medicine* held a debate for both sides of the G-spot issue, concluding that further evidence is needed to validate the existence of he G-spot. In 2012, scholars Kilchevsky, Vardi, Lowenstein and Gruenwald stated in the journal, "Reports in the public media would lead one to believe the G-spot is a well-characterized entity capable of providing extreme sexual stimulation, yet this is far from the truth." The authors cited that dozens of trials have attempted to confirm the existence of a G-spot using surveys, pathologic specimens, various imaging modalities and biochemical markers, and concluded:

> The surveys found that majority of women believe a G-spot actually exists, although not all the women who believed in it were able to locate it. Attempts to characterize vaginal innervation have shown some differences in nerve distribution across the vagina, although the findings have not proven to be universally reproducible. Furthermore, radiographic studies have been unable to demonstrate a unique entity, other than the clitoris, whose direct stimulation leads to vaginal

orgasm. Objective measures have failed to provide strong and consistent evidence for the existence of an anatomical site that could be related to the famed G-spot. However, reliable reports and anecdotal testimonials of the existence of a highly sensitive areas in the distal interior vaginal wall raise the question of whether enough investigative modalities have been implemented in the search for the G-spot.

A 2014 review from *Nature Reviews Urology* reported that "no single structure consistent with a G-spot has been identified."

History

The release of fluids had been seen by medical practitioners as beneficial to health. Within this context, various methods were used over the centuries to release "female seed" (via vaginal lubrication of female ejaculation) as a treatment for *suffocation ex semine retento* (suffocation of the womb), female hysteria or green sickness. Methods included a midwife rubbing the walls of the vagina or insertion off the penis of penis-shaped objects into the vagina. In the book *History of V*, Catherine Blackledge lists old terms for which she believes to be the female prostate (the Skene's gland), including *the little stream*, *the black pearl* and *palace of yin* in China, *the skin of the earthworm* in Japan and *saspanda nadi* in the India sex manual *Ananga Ranga*.

The 17th-century Dutch physician Regnier de Graff described female ejaculation and referred to an erogenous zone in the vagina that he linked as homologous with the male prostate; this zone was later reported by the German gynecologist Ernst Grafenberg. Coinage of the term G-spot has been credited to Addiego et al. in 1981 named after Grafenberg, and to Alice Kahn Ladas and Beverly Whipple et al. in 1982. Grafenberg's 1940s research, however, was dedicated to urethral stimulation; Grafenberg stated, "An erotic zone could be demonstrated on the anterior wall of the vagina along the course of the urethra." The concept of the G-spot entered popular culture with the 1982 publication of *The G Spot and Other Recent Discoveries About Human Sexuality* by Ladas, Whipple and Perry, but it was criticized immediately by gynecologists: some of them denied its existence as the absence of arousal made it less likely to observe, and autopsy studies did not report it.

—G-Spot entry
Wikipedia.org

~ 11 ~

Bare down there?

Should women shave their pubic hair?

Increasingly, more and more women are shaving down there: for two reasons—cultural and personal. Personal and cultural.

And even a discussion of this once-forbidden topic is now mainstream—discussed commonly and openly in the media.

Even the highly respectable *Atlantic* magazine has published a major article about *bare down there*.

The article "The New full-Frontal: Has Pubic Hair in America Gone Extinct?" by Ashley Fetters, was published Dec. 13, 2011.

She wrote:

The most staggering aspect of the bald-vulva phenomenon is just how quickly women(and men) have embraced it.

Less than two decades ago, the idea of 'taking it all off" seemed painful, unnecessary, and even vaguely fetishistic; as recently as 1996, one harrowing, particularly memorable vignette from Eve Ensler's groundbreaking play *The Vagina Monologues* effectively turned the idea of removing pubic hair at the request of a sexual partner into something cringle-worthy. And perverted. Trimming away a few stray strands during swimsuit season was one thing, but removing all the hair from one's genitals, effectively turning back the clock on puberty? Traumatizing. Selfish. Inhumane, even.

The character Carrie Bradshaw, in the television series *Sex in the City*, used a

Brazilian wax treatment to become bare and became liberated by it, Fetters says.

Victoria Beckham announced that she thought Brazilian waxes should be compulsory on all girls at age 15 and above; actress Eva Longoria said the same thing; that a Brazilian wax treatment to become bare was something every woman should try at least once.

Although sex, hygiene and clothing are all contributing factors … there's one main driving force behind America's villainization of public hair: pornography.

And Hugh Hefner's *Playboy* magazine led the trend for years. And the porn industry.

Now, Fetters says, quoting Ohio University media and culture Professor Joseph Slade, "the practice is wide spread in video porn today. Enough so that backlash has created a niche fetish for 'full bushes.'"

In an article "Pubic Shaving: Which Women? And Why?" in *Psychology Today*, Sept 15, 2015, Michael Castleman makes the point (as have others):

In ancient Greek, Egyptian and India art, some female nudes sport trimmed or shaved pubic hair. In Renaissance Italian art, female nudes were often depicted bald between the legs, but the art of the same era in Northern Europe typically showed full bushes. We don't know whether the Italian artists reproduced what they saw or indulged in artistic license.

Castleman cites an Indiana University study of what age, who shaves and how much:

In an internet -based survey, Indiana University researchers asked 2,451 women age 18 to 68 how they present their pubes. The results:

Full bush (nothing removed)

    18-24 12 %

    25-25 16 %

    30-39 19%

    40-49 28 %

    50 + 52 %

Trimmed with scissors

    18-24 29 %

    25-29 39 %

    30-38 50 %

    40-49 50 %

    50 + 37 %

Some removal (shaving, waxing, electrolysis)

    18-24 38 %

    25-29 32 %

30-39 23 %

40-49 16 %

50 + 9 %

Bald (no hair)

18-24 21 %

25-29 12 %

30-39 9 %

40-49 7 %

50 + 2 %

Castleman writes: "Pubic hair removal is clearly age-related. The younger the woman, the more likely she is to tinker with her presentation."

What are the health risks related to removing pubic hair? Minor at best: razor burns, or burns from using hot wax (Brazilian waxing); allergies to depilatories, creams other products; cuts; infections from unclean equipment; blisters; pimples, redness, itching.

Ultimately the key word for many women is: *empowerment.*

In the website Nerve.com, June 20, 2014, under the title "How I Found Empowerment in Letting a Man Shave My Pubic Hair," Jennie Gruber wrote that her boyfriend asked to shave her:

So I decided to be game. I decided to let him shave me.

And that was how I ended up in our windowless bathroom, watching him groom me, as my precious bush fell in little clumps to the blue bathmat.

And when I looked down, I saw something I never expected.

I saw my vulva with a clarity I never imagined … It was as if my lips and clit had been hiding from me. I was really beautiful … I experienced first hand how preconceived notions about sexuality and the body can change completely ….

My experience being shaved by a partner taught me that it was juvenile to think that *not* shaving was inherently empowering, or that being "unfeminine" was empowering. It *is* empowering to consciously make choices based on our own personal identity. And being open to new possibilities.

~ 12 ~

Her period

Every women in the world has monthly periods, beginning, in the U.S. around age 12, and ending with menopause between ages 45 to 55.

One of the clearest explanations is from the website www.medicinenet by Charles Patrick Davis, MD, PhD:

> Menstruation (men-STRAY-shuhn) is a woman's monthly bleeding.

> ✶ ✶ ✶

> Menstration is a monthly shedding of a female's uteral living; it lasts about 3 to 5 days (average) and contains blood and tissue that exits her

body through the cervix and vagina—the first day of menstruation is the first day of your period.

The menstral cycle is the recurrent approximately monthly menstruation.

The menstral cycle is the hormone driven cycle; day 1 is the first day of your period (bleeding) while day 14 is the approximate day you ovulate and if an egg is not fertilized, hormone levels eventually drop and at about day 25, the egg begins to dissolve and the cycle begins again with the period at about day 30.

Most periods vary somewhat; the flow may be light, moderate or heavy and can vary in length from about 2 to 7 days; with age, the cycle usually shortens and becomes more regular.

The internet dictionary www.wikipedia.org states:

The first period usually begins between twelve and fifteen years of age, a point in time known as menarche. However, periods may occasionally start as young as eight years old and still be considered normal. The average age of the first period is generally later in the developing world, and earlier in the developed world. The typical length of time between the first day of one period and the first day of the next is 21 to 45days in young women, and 21 to 31days in adults (an average of 28 days). Menstruation stops occurring after meno-pause, which usually occurs between 45 and 55 years of age. Bleeding usually lasts 2 to 7 days.

The menstrual cycle occurs due to the rise and fall of hormones. The cycle results in the thickening of the lining of the uterus, and the growth of an egg (which is required for pregnancy). The egg is released from an ovary around day fourteen in the cycle; the thickened lining of the uterus provides nutrients to an embryo after implantation. If pregnancy does not occur, the lining is released in what is known as menstruation.

And, Wikipedia states:

Up to 80% of women report having some symptoms prior to menstruation. Common signs and symptoms include acne, tender breasts, bloating, feeling tired, irritability and mood changes. These may interfere with normal life, therefore qualifying as premenstrual syndrome, in 20 to 30% of women. In 3 to 8% symptoms are severe.

Premenstrual syndrome is commonly known, and commonly abbreviated, PMS.

And menopause can result in: night sweats; hot flashes; insomnia; moodiness; irritability; reduced sex drive; pain during intercourse.

In a remarkable recent exchange, Republican presidential candidate Donald Trump appeared to accuse—on national television—Fox News anchor Megyn Kelly of having her period.

The Washington Post carried this headline Saturday August 8, 2015, and article written by Philip Rucker:

Trump says Fox's Megyn Kelly had 'blood coming out of her whatever'

Republican presidential candidate Donald Trump said Friday that Fox News Channel anchor Megyn Kelly "had blood coming out of her eyes " when she aggressive-ly questioned him during Thursday's presidential debate.

> "She gets out and she starts asking me all sorts of ridiculous questions," Trump said in a CNN interview. "You could see there was blood coming out of her eyes, blood coming out of her wherever. In my opinion, she was off base."

> (On Saturday morning, Trump tweeted that he was referring to Kelly's nose. His campaign also issued a statement, claiming Trump said "whatever" instead of "wherever,"

while again repeating that the reference was to her nose.)

In Thursday's debate, Kelly questioned Trump over his history of offensive statements about women.

Internet medical websites carry extensive articles about menstruation. Planned Parenthood also offers consultations about menstruation and menopause; doctors and nurses can also offer advice.

13

End Note ...

Any further questions, gentlemen ... ?

About the author ...

Lander Moore is the pseudonym of a veteran non-fiction writer; he admits he is as mystified about women as any other man in the world.

www.ingramcontent.com/pod-product-compliance
Lightning Source LLC
Chambersburg PA
CBHW071139280326
41935CB00010B/1298